Preeni Fernando
Ranjith Premasiri

Analysis of Potato Productivity in Sri Lanka

AF138617

Preeni Fernando
Ranjith Premasiri

Analysis of Potato Productivity in Sri Lanka

LAP LAMBERT Academic Publishing

Impressum / Imprint

Bibliografische Information der Deutschen Nationalbibliothek: Die Deutsche Nationalbibliothek verzeichnet diese Publikation in der Deutschen Nationalbibliografie; detaillierte bibliografische Daten sind im Internet über http://dnb.d-nb.de abrufbar.
Alle in diesem Buch genannten Marken und Produktnamen unterliegen warenzeichen-, marken- oder patentrechtlichem Schutz bzw. sind Warenzeichen oder eingetragene Warenzeichen der jeweiligen Inhaber. Die Wiedergabe von Marken, Produktnamen, Gebrauchsnamen, Handelsnamen, Warenbezeichnungen u.s.w. in diesem Werk berechtigt auch ohne besondere Kennzeichnung nicht zu der Annahme, dass solche Namen im Sinne der Warenzeichen- und Markenschutzgesetzgebung als frei zu betrachten wären und daher von jedermann benutzt werden dürften.

Bibliographic information published by the Deutsche Nationalbibliothek: The Deutsche Nationalbibliothek lists this publication in the Deutsche Nationalbibliografie; detailed bibliographic data are available in the Internet at http://dnb.d-nb.de.
Any brand names and product names mentioned in this book are subject to trademark, brand or patent protection and are trademarks or registered trademarks of their respective holders. The use of brand names, product names, common names, trade names, product descriptions etc. even without a particular marking in this work is in no way to be construed to mean that such names may be regarded as unrestricted in respect of trademark and brand protection legislation and could thus be used by anyone.

Coverbild / Cover image: www.ingimage.com

Verlag / Publisher:
LAP LAMBERT Academic Publishing
ist ein Imprint der / is a trademark of
OmniScriptum GmbH & Co. KG
Heinrich-Böcking-Str. 6-8, 66121 Saarbrücken, Deutschland / Germany
Email: info@lap-publishing.com

Herstellung: siehe letzte Seite /
Printed at: see last page
ISBN: 978-3-659-69399-1

Copyright © 2015 OmniScriptum GmbH & Co. KG
Alle Rechte vorbehalten. / All rights reserved. Saarbrücken 2015

CONTENTS

1. INTRODUCTION

1.1 Potato

Potato (*Solanum tuberosum* L) is a cool-season vegetable that ranks with wheat and rice as one of the most important staple crops in human diet around the world. Potatoes are not roots but specialized underground storage stems called "tubers". Maximum tuber formation occurs at soil temperatures between 13^0 C and 21^0 C. The tubers fail to form when the soil temperature reaches 27 ^{0}C. Potatoes can withstand light frost and can be grown most of the countries in cooler regions.

1.2 Production Areas in Sri Lanka

Potato is the most popular crop of up-country farmers due to its high net return. At present potato is extensively cultivated in highlands of Nuwara Eliya district and midlands of Badulla district.

Nuwara Eliya District

The major potato growing season in the district is, "*yala*" (February –May) and minor season is "*maha*" (September – December). Planting is avoided during May-July because of heavy wind and rain, and in December and January due to night frost. In this higher location imported seed potato is used to produce February crop while for the second crop, June produced tubers are used for planting. Currently about 1490   ha are devoted to potato cultivation.

Badulla District

Potato is also widely cultivated in Badulla district during "*yala*" and "*maha*" seasons. In "*yala*" (May–August) potatoes are grown in paddy fields following rice harvest while "*maha*" (November –February) in highlands. May planted crop is

1

grown from February produced tubers and harvested in August-September. At present there is about 4150 ha of land under potato cultivation.

Low Country Dry Zone

Kalpitiya in Puttalam district and Jaffna district are the other two districts where the potatoes are planted to a lesser extent during November and December, immediately after heavy rains. Potatoes are harvested in February before the temperature rises.

Nature has gifted Sri Lanka with immense edaphic and climatic resources for ideal production of potatoes.

1.3 The Environment of Potato Growing Areas in Sri Lanka

The study area (Central hill country) falls within intermediate and wet climatic zones of Sri Lanka (Figure 1.1) which receives an average annual rainfall of 2500-1750 mm and 5000-2500 mm respectively (Figure 1.2). Mean annual temperature of the study area belongs to two zones such as 15-17.5^{0}C and >15^{0}C (Figure 1.3). Agro-ecological regions of the study area are up-country wet zone (1,2,3) and up-country intermediate zone (1,2,3) (Figure 1.4). Geomorphological features of the area are mainly high elevated ridges and mountain range, plateau and undulating plains and basinal structures. Examples are Uva Basin, Pidurutalagala ridge, Horton Plain, High Plain and Idalgashinna-Ohiya ridges (Figure 1.5). The relief of the area is over 900 m (Figure 1.6). Soil types, climate and cropping seasons for potatoes are shown in Table 1.1.

2

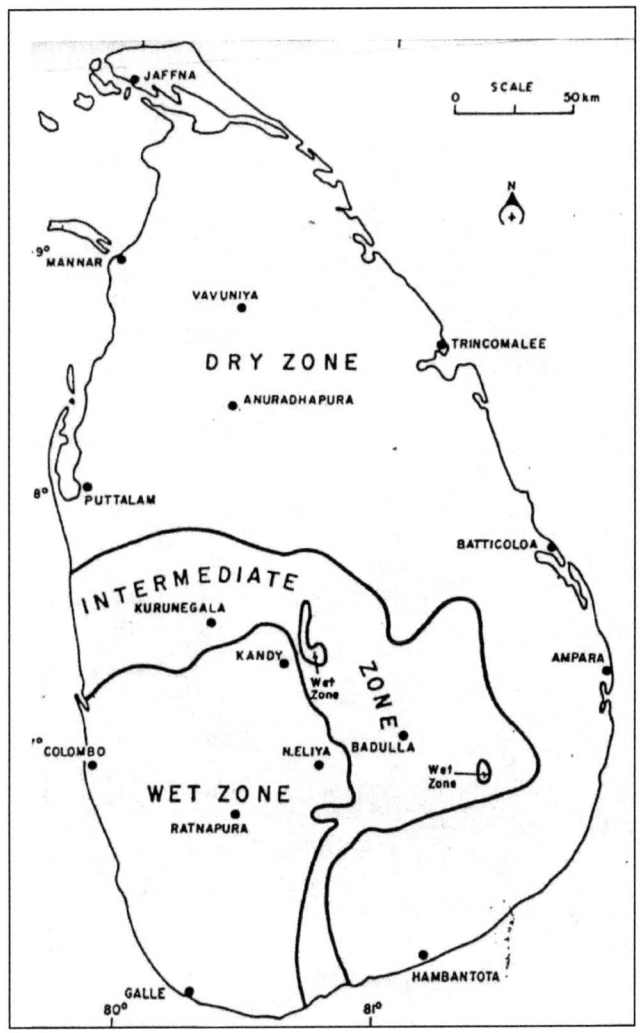

Figure 1.1. Demarcation of Wet, Dry and Intermediate Zones.

(Source: C.R.Panabokke, 1996)

3

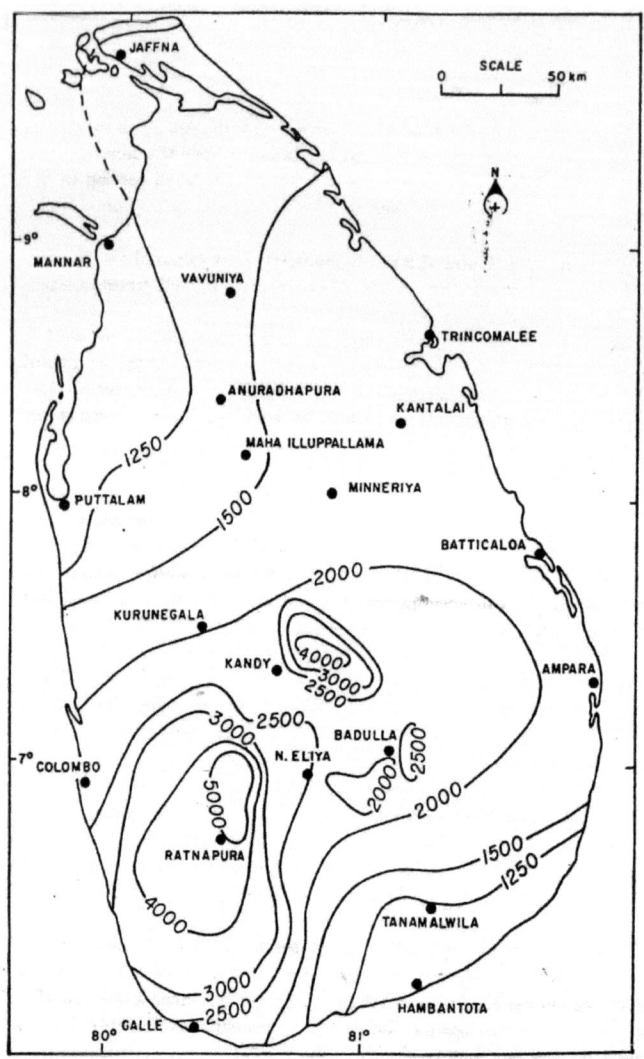

Figure 1.2. Mean Annual Rainfall (mm)

(Source: M.Domros, 1977, cited by Panabokke,1996)

4

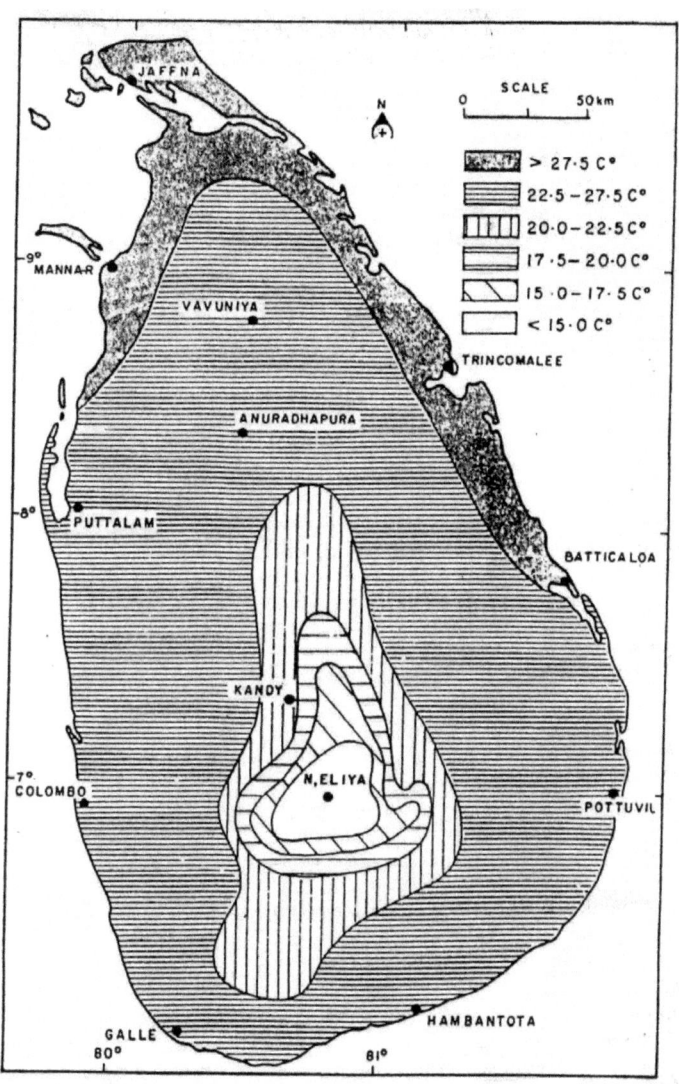

Figure 1.3. Mean Annual Temperature

      (Source: R.Kannagara, 1982,cited by Panabokke,1996)

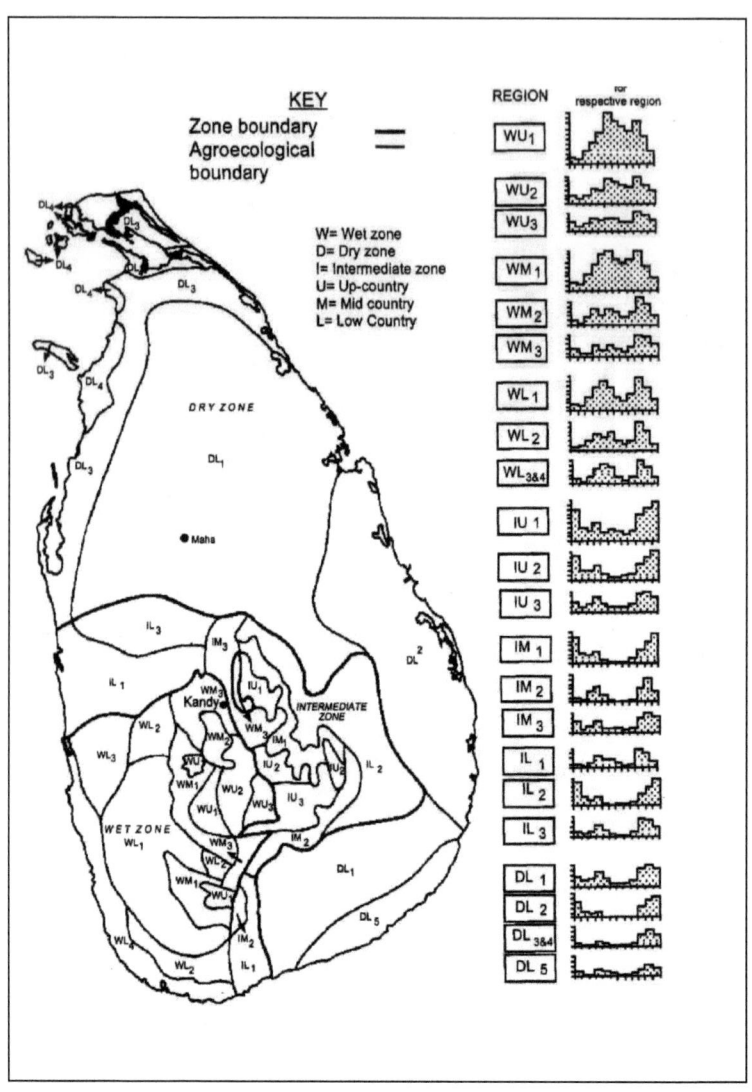

Figure 1.4. Agroecological Map of Sri Lanka (Source:
C.R.Panabokke,1996)

6

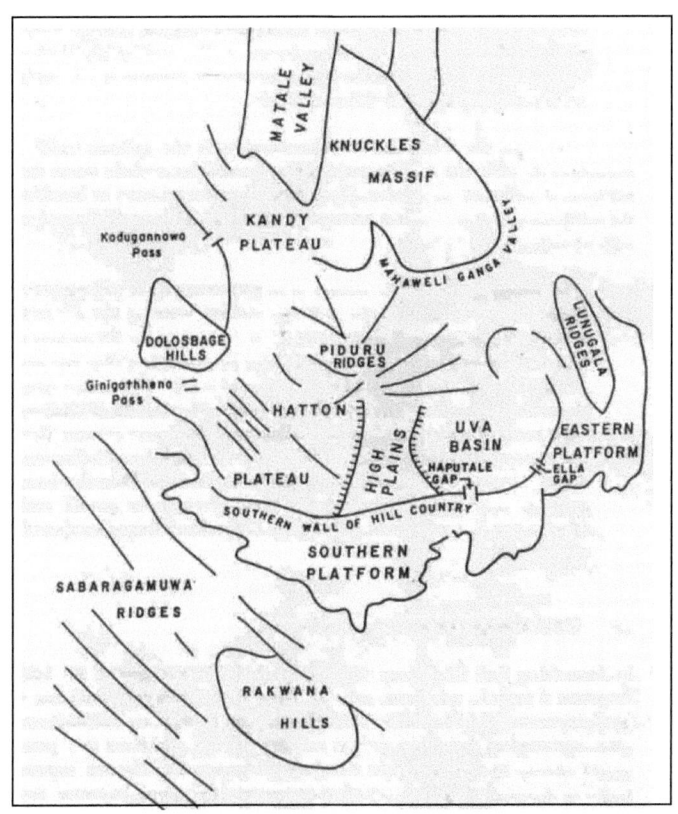

Figure 1.5. Physiological Regions of the Hill Country
(Source: E.Cook, 1931,cited by Panabokke,1996)

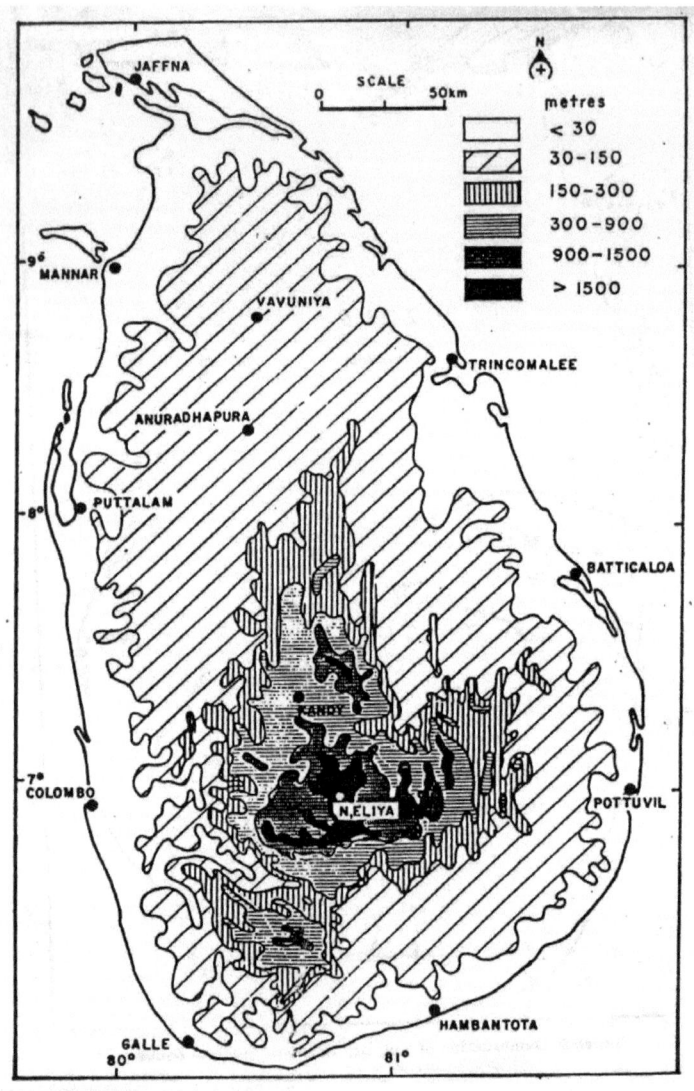

Figure 1.6. Generalized Relief Map of Sri Lanka

(Source: C.R.Panabokke, 1996)

8

Table 1.1. The environment of potato growing areas (Nuwara Eliya and Badulla districts)

Area	Altitude (m)	Agro-ecological zone	Soil type	Potato growing season	Mean daily temp. range Min.	Max.	Average annual rainfall (mm)
Nuwara Eliya	1200-2000	IU2 WU2 Wu3	Red-yellow Podzols, mountain regosols	Feb - May Sept. - Dec.	8 -12 10 -12	21- 22 19- 20	2800
Badulla	900 - 1200	IU3, IU2	Red-yellow podsols	May - Aug. Nov. - Feb.	17 -18 15 -16	25 -27 22 - 25	2300

1.4 Soils and Fertilization

Potatoes perform well on a wide range of soils, but the most suitable soil is deep, well-drained sandy loam or silty loam (Crop profile-Potato). Even though the root system of potato can grow down about 60 cm, roots cannot absorb sufficient water from fields which do not posses good water holding capacity. Applying organic matter in the form of well-rotten manure, compost or similar material will improve the soil structure for better water holding capacity.

9

Well-drained latosols, regosols and non-calcic brown soils of the low country dry zone and all up-country soils of Sri Lanka are preferable but ill-drained soils are generally unsuitable. The optimum soil pH range is about 5.5-6.0.

1.5 Seed Flow

The possible seed source for *"yala"* planting in Nuwara Eliya area is imported seed or early *"maha"* seed. Potatoes harvested in *"yala"* season in Nuwara Eliya are still physiologically rather young when planted in the second planting season (*"maha"*) in Nuwara Eliya but are at a good physiological stage when planted in Badulla and Jaffna *"maha"* season. For *"yala"* season in Badulla, the seed comes either from harvest of *"maha"* season, Nuwara Eliya or harvest of *"yala"* season, Badulla.

1.6 Recommended Potato Varieties for Sri Lanka

Potato Variety	Skin Colour	Flesh Colour	Disease Resistance	Potential Yield
Desiree	Red	Pale yellow	Resistant to tuber blight,	15-20 t/ha
Sante	Yellow	Pale yellow	Resistant to late blight	15-20 t/ha
Raja	Red	Yellow	Resistant to tuber blight, leaf roll Virus A and Virus X	15-20 t/ha
Granola	Yellow	Yellow	Resistant to leaf blight, moderate resistance to Virus	20-25 t/ha

			A and Virus Yn, resistant to golden nematode A	
Kondor	Red	Pale yellow	Resistant to late blight, tuber blight, Virus X and Virus A.	20-25 t/ha
Isna	Yellow	Yellow	Resistant to late blight	20 t/ha
Sita	Yellow	White	Tolerant to late blight	20-25 t/ha
Krushi	Yellow	White	Tolerant to late blight	20 t/ha
Lukshmi	White	white	Moderately resistant to late blight	20 -25 t/ha
Manike	White	white	Moderately resistant to late blight	20-25 t/ha
Hillstar	Cream	Yellowish white	Resistance to late blight	25 t/ha

(Source: Department of Agriculture, Sri Lanka,2006)

1.7 Production

Potato is a highly profitable crop for farmers in Nuwara Eliya and Badulla districts. However, average productivity of potato in Sri Lanka is about 12 t/ha and is very low compared to other potato growing countries in the world. The cost of production is also very high, since seed potatoes are very expensive (Rs.80-100/kg) and accounts for more than 50% of the total cost of production (Nugaliyadde, 2003).

Procurement of good quality seed at a reasonable price is one of the chief complaints of the growers. The imported seed is preferred because it yields more (1:11) than crops grown from their own seed (1:9). These yields are far below the world average which usually generates 20-25 kg of marketable tubers from each kilogram of seed planted (Annon, 1993).

The national average of potato productivity remained at 12 t/ha even though attempts have been taken to increase the productivity at different capacities. The production is not sufficient to meet the demand of the country. Therefore, research and development activities on potato should be focused to increase the extent and productivity. In addition emphasis should be given to reduce the cost of production. The productivity is closely associated with the adoption of technology.

Productivity is determined by the trend in adoption of technology such as selecting high yielding and disease resistant varieties, irrigation methods, fertilizer and pest and disease management practices. Poor adoption of technology increases the price for inputs and decreases the price for the product.

1.8 Application of Geographic Information System for Evaluation of Potato Cultivation in Sri Lanka

Land evaluation is a basic technique and often used for land-use planning and for estimating land productivity for a selected crop. Different empirical modelling approaches to predict land productivity for crops under the wide range of weather and soil conditions have been described (Pratummintra *et al.*, 2002). Most of these models are designed to use available climatic and soil information as statistical average and generalized crop phenology. A production situation is a hypothetical land-use system, with one or only a few relevant land characteristics and/or land qualities and the production calculated is not the actual but potential production.

However, modelling and handling of geographic data for decision making is very important in any aspect. Especially in agricultural sector, if there is a user friendly method for decision making it is the most worthwhile exercise. Therefore, Geographical Information System (GIS) plays a remarkable role in agriculture sector. GIS is a set of computer tools for collecting, storing, retrieving, transforming and displaying spatial data from the real world. GIS becomes quite popular and its use in all countries is still growing. Handling the variability in climatic conditions and soil information can be approached through the use of GIS. Using GIS techniques, it is possible to produce thematic maps as an output with information on the impact of differences in climatic and soil on land productivity for a specific crop.

Even in Sri Lanka, potato cultivation is restricted to a very few districts and to small clusters, it is not easy to maintain a database and modelling them. However, in this study it was aimed to use GIS as a modern computer based powerful tool for modelling potato cultivation in Sri Lanka in order to access the land, environment and varietal suitability of potato in Nuwara Eliya and Badulla districts.

All factors which are important for potato cultivation are studied and analysed to determine the significant effect on the production.

1.9 Objectives

- To identify high yielding varieties/hybrids which are adaptable to local environment.
- To introduce farmer awareness approach to increase the potato production system using GIS based model.

2. LITERATURE SURVEY

2.1 History and Production Trends

The successful cultivation of potatoes was first recorded in Sri Lanka in1812 when it was introduced to Morawak Korala. According to Abeyaratne (1985) potatoes were probably introduced by Dutch who ruled the coastal areas from 1656 to 1792. It is reported that all attempts at growing potatoes in the centre of the potato producing area were done by Samuel Baker around 1850. However, it was only in 1909 that the local potatoes of Uruguay origin were grown at Hakgala Gardens. Later some British planters in up-country cultivated potatoes in their home gardens for their own use. Meanwhile the Department of Agriculture was concerned with potato cultivation for a considerable period (Mahakumbura, 1980).

In 1948, the Department of Agriculture began working on potatoes, devoting the first few years investigating the cultivars suitable for growing, cultural practices, fertilizer applications and control of pests and diseases. The cultivation of potatoes on a large scale was started in 1951-1952, but it failed due to various reasons. Cultivation was again attempted in 1957 in Badulla and Nuwara Eliya districts, but the farmers were not able to get a reasonable income because imported tablestock potatoes were available in the local market at a very low price. When the government stopped the import of tablestock potatoes in 1967 to encourage local production, the farmers once again started to cultivate this crop. This strategy was immediately successful and there was a rapid expansion of the acreage of potatoes (Table 2.1) (Sathiamoothy et al. 1985).

Table 2. Potato Production of Sri Lanka (1961 – 1982)

Year	Area cultivated (ha)	Production (Tons)	Yield (T/ha)
1965	795	4732	5.95
1967	1338	11533	8.62
1969	2934	26984	9.20
1975	3086	27983	9.06
1977	3062	29164	9.52
1979	3992	46950	11.76
1980	5154	64779	12.57
1982	5058	81000	16.01

(Source: Adopted from Sathamoorthy, *et al.*, 1985)

2.2 States of Potato Cultivation in Sri Lanka

Until mid 1990s, potato was the most popular crop among the farming community in up-country area due to its high biome production and high net return. However, the low domestic production and high market price limit the availability of potatoes for all classes of people, and this is attributed to low per capita consumption. Importation of large quantities of consumption potatoes under the open economy led to reduce the extent under cultivation drastically from 1998. However, latter part of the year 2000 the Government imposed a specific tax of $ 0.25 per kilo of imported potatoes ensuring a high price for local potatoes and increasing domestic production (Kularatne, 2003).

2.2.1 Extent and production

Mazeen *et al.* (2002) reported that there was 9000 ha of land under potato cultivation in Sri Lanka, mainly in Badulla and Nuwara Eliya districts and lesser extent in lower elevations. The lowest extent and production were recorded in year 1999 and there was a slight increase in extent and production, 5747 ha and 70069 t respectively in year 2001, of which 4159 ha from Badulla and 1489 ha from Nuwara Eliya districts and national productivity remained at 12 t/ha (Table 2.2).

Table 2.2. Area, Production and Yields of potatoes grown in Sri Lanka (1993-2001).

Year	Area (ha)	Production (t)	Yield (t/ha)
1993	7000	78000	11.1
1994	8480	70590	9.5
1995	9025	108000	12.8
1996	7925	100755	12.7
1997	6469	66484	10.3
1998	2327	25399	11.1
1999	2171	27171	12.5
2000	3646	48409	13.3
20001	5747	70069	12.2

(Source: Department of Agriculture, Sri Lanka)

2.3 Potato Varieties

2.3.1 Recommended varieties

Desiree, Hilstar and other suitable varieties are Granola, Raja, Isna, Kondor and Lyra. Desiree is the only popular variety among the farmers in Jaffna and Kalpitiya.

2.3.2 Varieties have been imported as seed potatoes during the year 2004/2005

Matador	Murato	Vigro	Alurera
Cycloon	Teragold	Premier – E	Claret
Lady Balfour	Eve Balfour	Sinora	SA 9 – 030 – 03
Octavia	Baltica	Red lady	Donella
Ajiba	Bellini	Rodeo	Bartina
Asterix	Dura	Binella	Raja
Artemis	arinda	Arnova	Baraka
Aladin	Donella	Miranda	YP 97-064
KR 195-017	Secura	Finka	Laura
Lyra	Milva	Bellarosa	1755 –94
Florence	Delphine	Denis	Alwara
Chantal	Ranpa	Carola	Rikea
Desiree	Brrnadette	Vivaldi	Fabula
Derby	Caesar	Cisero	Granola
Kondor	Santae	Provento	Ultra
Secura	Satina	Marabe	

2.3 3 Varieties imported as seed potatoes in 2005/06 (until first week of January).

Margarita	Safrane	Alaska	Desiree
Granola	Dura	Ultra	Binella

18

Provento	Arnova	Lyra	Jelly
Marabel	Laura	Vigro	Murato
Kuroda	Arinda	Raja	Aladin
Kondor	Gala	Beluga	Vivaldi
Rodeo	Derby	Cicero	Asteria
Caesar	Bernadette	Aida	Atlas
Rosanna	Artemis	Andante	Princess

(Source: Potato Reports 2004/05 and 2005/06 at National Plant Quarantine Service)

2.4 Time of Planting

Season	Location	Planting date
"Maha"	Nuwara Eliya	August-September
	Badulla	November-December
	Jaffna & Puttalum	Mid-Nov.-Mid December
	Kalpitiya	Mid-Oct.-Mid December
"Yala"	Nuwara Eliya	February-March
	Badulla (rich land)	July-August

(Source: Department of Agriculture, Sri Lanka,2006)

2.5 Planting of Potato

2.5.1 Seed selection

Disease free well sprouted tubers of 28-55 mm diameter from a reliable source should be used. The tubers must also be free from varietal mixing. Generally, potatoes are started from "seed pieces" rather than true seeds. Small whole tubers are sometimes used to establish the field planting, but more commonly tubers are cut into large chunky pieces of 28-42 g that bearing at least one good eye. The pieces should either be immediately planted or healed by holding at 18°C and high humidity for 1 to 2 days after cutting. Healing will reduce rotting.

In commercial production seed pieces are treated with dust formulated fungicides immediately after cutting. Each tuber should have 4-5 sprout of 2-3 cm long with more visible root growing point. Physiological age of the seed pieces will affect the size and number of tubers. Old potato seed is undesirable because it produces more stems and large number of small potatoes than young seed.

2.5.2 Seed rate

Generally, farmers keep their own seed potato harvested from "*maha*" season for the following planting season "*yala*" and that is around 50% of the total requirement. It has been reported that seed rate for consumption potato production is 2500 kg/ha whereas the rate for seed potato production is 4000 kg/ha.

2.5.3 Planting

Seed tubers are placed 25 cm apart along the furrow *ie* between raw 60 cm and within raw 25 cm and close with soil up to 6-10 cm height. Within raw spacing can be reduced if seed potatoes are less than 28-55 mm grade. As main stems are the best unit of population in potato production, spacing can be adjusted to get optimum stem density of 30-35 stem/m^2. For seed production potatoes are planted in close spacing of 45 cm x 20 cm instead of 60 cm x 25 cm. Furrow planting is recommended for very well-drained soils.

2.5.4 Earth-up and weed control

Immediately after planting pre-emergence weedicides, such as Metribuzin (not for sandy soils) can be sprayed. Sprouts are liable to be affected if sprayed after emergence. Usually, earth-up is done 3-4 weeks after planting. A good earth-up effectively controls the weeds.

2.5.5 Roughing
Diseased and off-type plants are removed regularly.

2.6 Fertilizer Application

2.6.1 Fertilization

It is reported that harvesting one ton of potatoes removes 3.6-6.8 kg (8.0-15.0 pounds) of N, 0.72-3.1 kg (1.6 –7.0 pounds) of P_2O_5 and 4.5- 11.8 kg (10.0-26.0 pounds) of
K_2O from the soil. Potato is identified as a heavy feeder in terms of phosphorous and potassium, but nitrogen must be carefully managed to provide adequate but not excessive amounts. As little as 11.3 kg (25 pounds) of nitrogen at planting is often recommended as the plant does not use much nitrogen in its first 4-5 weeks of growth. A later side-dressing of 9.0-18.0 kg (20-40 pounds) nitrogen is sometimes applied after shoots emerge but before they are 10-15 cm high. Excess fertilizer increases salt accumulation which decreases crop growth (Cop profile –Potato).

2.6.2 Organic fertilizers

Well-rotten cow dung or compost at the rate of 10-20 t/ha or 5-10 t/ha of poultry manure is applied to the furrows before planting.

2.6.3 Chemical fertilizer

1. Nuwara Eliya District

   N-200 kg/ha, P_2O_5-115 kg/ha, K_2O-90 kg/ha, on sandy regosols, N and K can be split applied in 3-4 doses.

   a)　Basal: The following formulations and rates are applied.
   - i)　Ammonium sulphate　-　500 kg/ha
   - ii)　TSP　-　259 kg/ha
   - iii)　Muriate of potash　-　75 kg/ha

   b)　Top dressing – apply at earth-up
   - i)　Ammonium sulphate　-　500 kg/ha
   - ii)　Muriate of potash　-　75 kg/ha

2. All other areas: N- 100 kg/ha, P_2O_5- 115 kg/ha, K_2O- 90 kg/ha.

   a)　Basal –the following formulations and rates are applied.
   - i)　Ammonium sulphate　-　250 kg/ha
   - ii)　TSP　-　250 kg/ha
   - iii)　Muriate of potash　-　75 kg/ha

   b)
   - I)　Ammonium sulphate　-　250 kg/ha
   - ii)　Muriate of potash　-　75 kg/ha

2.7 Irrigation

Irrigation must be done to keep the soil moist. Regosols usually require irrigation daily. With other soil types, irrigation every 3-4 days may be sufficient.

2.8 Harvesting

Harvesting can be done when 80% of the leaves become yellowish. In the up-country maturity of the tuber can be hastened by cutting haulms (stems) about 10 days before the expected date of harvesting. Tuber maturity can be checked by rubbing with thumb. Resistance of peeling when rubbed indicates the maturity.

Tubers should be handled carefully to avoid mechanical damage. All damaged and diseased tubers should be removed before storing. Yield of 15-25 t/ha under good management was recorded.

2.9 Pests and Diseases of Potato

The pests and diseases cause economic losses to growers, also results in increased prices of produce to consumers. It has been estimated that 11.8% of the total food produced in the world is lost due to diseases before the harvest of the crop. With regard to potato 21.8% of potato crop is lost due to diseases (Babu, 1999).

2.9.1 Pests of Potato

- Vegetable leaf miner (*Liriomyza huidobrensis*)

The pea leaf miner is a highly polyphagous leaf miner capable of causing severe damage to crops. Affected crops include field and glasshouse grown vegetables, flowers and potatoes. This pest was first arrived in Sri Lanka through imported flower plants and spread rapidly in Nuwara Eliya region. This happened due to intensive cultivation pattern of the region and lack of natural enemies for the new pest. *L. huidobrensis* is apparently indigenous to cooler, mostly highland areas in the country.

Larvae are usually associated with the midrib and lateral veins. A mine usually begins on the upper leaf surface and moves to the lower surface after a few millimetres of feeding by the larva. Adults punch leaves for both feeding and oviposition. Punctures and mines may be numerous enough to reduce photosynthesis greatly and may kill young plants.

- Golden Cyst nematodes (*Globodera rostochiensis*)

The Golden Cyst nematode is a root invading nematode capable of creating losses up to 70%. This was first introduced to Sri Lanka in 1979 through imported seed potatoes and seventy five percent of the farmers in Nuwara Eliya region were severely affected by the nematode.

Nematodes remove nutrients from the roots and diminish supply of nutrients and water to the stems and leaves by injuring the roots and stunting their growth. Moderately infested plants normally produce smaller tubers, but infestation does not affect tuber number. Heavily infested plants grow and develop poorly. As temperature becomes warmer wilting of plants occurs, tuber formation becomes poor or tubers are either absent or small.

- Potato Tuber Moth (*Phthorimaea operculella*)

This moth attacks potato in two ways.

>Leaf mining: Larvae tunnel the leaf veins and stems. This results in the loss of leaf tissues, death of growing points and breakage of stems.
>Tuber infestation: Damage occurs in the field and in storage.

- Cutworm (*Agrotis* spp. and other *Noctuidae* spp.)

They are the larvae of several noctuid moth species that cut through the stems of young plants. During first three weeks is likely cutworm damage.

- Green peach aphids (*Myzus persicae*)

Aphids are carriers of virus and harbouring on volunteer plants around the field. Direct feeding damage is not significant.

- Mites (*Terranychus* sp.)

They develop on the crop during warm weather.

- White grubs (*Melonontha* spp. and *Anomala* spp.)

They are larvae of scarabaeid beetles. The larvae damage roots, stems and tubers. Market quality of tubers is reduced by cavities.

- Root eating ants (*Dorylus orientalis*)

In dry weather they do considerable damage to tubers and stems. Adequate irrigation during dry weather will reduce the damage.

- Leafhoppers

Leafhoppers migrate from one area to another and fly away in large numbers when foliage is disturbed. Leafhopper damage includes curled or wrinkled foliage and "hopper burn" (caused by leafhoppers feeding indicated by brown edges on leaves)

(IPM programme, 2003).

2.9.2 Fungal diseases

- Potato late blight (*Phytophthora infestans*)

This is known as the most serious disease in many potato growing regions in the world. It was reported in 1840 in Northern Europe and the devastating epidemic of late blight of potato in 1945-1946 in Ireland caused a widespread famine resulting in death of nearly one million people due to starvation and emigration of more than one and half million Irish people from Ireland to USA. Late blight is most destructive in areas with frequent cool, moist weather. Late blight may kill foliage and stems of potato any time during growing season. It can also attack tubers in either in the field or storage. Late blight may cause total destruction of all plants in a field within a week or two when weather is cool and moist.

- Stem canker and Black scurf (*Rhizoctonia solani*)

Stem canker and Black scurf of potato are caused by *Rhizoctonia solani*. These diseases occur wherever potato is grown and is serious in Germany, Netherlands and the USA. This fungus is present in nearly all soils because it has a wide host range, survives in plant debris and as sclerotia which is easily disseminated on tubers. This fungus is also very important as one of the most important organisms causing damping off of seedlings.

- Early Blight (*Alternaria solani*)

Unlike late blight early blight does not cause serious damage to potatoes. Early blight is widely distributed in all parts of the world. It occurs in warm dry seasons and reported in Badulla, Welimada, Bandarawela and Jaffna.

- Fusarium wilt and dry rot (*Fusarium oxysporum*)

These diseases are also distributed worldwide. Fusarium wilt mostly occurs in warm climates. Infection of roots may lead to decay of roots and tubers before plants are fully established. Since the fungus is soil-borne it first enters the roots and mother tubers. Vascular tissues of stems and tubers become browning resulting in discolouration of the vascular ring. It is a prominent symptom. Stolon attachments and eyes appear as circular brown rotted areas.

- Dry rot

Dry rot is one of the serious storage problems. It causes shrinkage of tubers and makes them unsuitable for consumption. The fungus enters the tubers through wounds at lifting from the soil.

2.9.3 Bacterial Diseases

- Bacterial wilt (*Ralstonia solanaceaarum*)

This disease is serious in warmer regions. It is widely distributed in tropical countries and partially serious in southern states of America, India, Australia and Sri Lanka. In Sri Lanka bacterial wilt occurs mostly in Badulla district.

- Black leg and soft rot (*Erwinia carotovora*)

Black leg causes a soft rot of the tubers and stems of potato in many parts of the world including Europe. In Sri Lanka it occurs in waterlogged fields in Nuwara Eliya. This disease is especially harmful in humid climates.

- Common Scab (*Streptomyces scabies*)

This disease is widely distributed in all potato growing regions of the world except where soils are very acidic. The losses caused by the disease have been estimated as several million dollars. Severe scab reduces yield by 15-20% due to lowering of the market grade of potatoes. Scabby potatoes have poor consumer demand as well as undesirable as seed.

2.9.4 Viral Diseases

- Leaf roll virus (PLRV)

This is the most destructive virus disease of potato and is common in all potato growing countries including Sri Lanka. Yield losses in highly susceptible varieties can reach up to 90%.

- Potato virus X (PVX)

This virus is known as Latent Mosaic Virus and may cause yield loss above 10%, and is transmitted through infected tubers or by sap either through hands of workers or through tools. Usually, the disease spreads by contact of diseased plants with healthy plants (Babu, 1999).

2.10 Marketing

After keeping their requirement of seed, the farmers in Nuwara Eliya and Badulla districts sell the balance of the harvest immediately to private dealers. The farmers in Jaffna and Puttalam sell their entire harvest as tablestock potatoes as their produce is not suitable for the seed. Most farmers prefer to sell their produce to the private dealers as they pay a higher price. The price of potato fluctuates widely throughout the year, because of the coincidence of harvest in Jaffna, Puttalam and Badulla in January, February and March. Hence, there is a drop in price during these months. Normally, farmers in Jaffna harvest their crop prematurely in order to get a better price and to have the land available for cultivation of the next crop.

Potatoes harvested from Jaffna have poor cooking and keeping quality. This tends to further decline the price during these months due to poor consumer acceptance. Another drop in price occurs in September and October as the farmers in Badulla harvest their "*yala*" crop. They also harvest premature tubers in order to compete in local markets. During June-July and December when only farmers in Nuwara Eliya harvest their potatoes, however tabltestock prices are high during this period because they sell most of their produce as seed to other districts (Sathiamoorthy, *et al.*, 1985).

28

3. MATERIALS AND METHODS

3.1 Field Survey

Field surveys were conducted in *"yala* and *"maha"* seasons, in 2004 and 2005 with 80 potato farmers in Nuwara Eliya and Badulla districts in up-country where potato is cultivated extensively and there were about 5747 ha of land under potato cultivation in these two districts. In total 80 farmers were interviewed by using a comprehensive questionnaire. Farmers were selected randomly talking to villages and looking at cultivated fields. Most fields in the study area were small and ranged from 0.1 – 0.2 ha.

The information was collected by interviewing farmers and inspecting their fields during growing seasons. Regular monitoring was done in selected fields in the study area by the Technical Assistant. Information on seed potato, variety, cropping system, time of planting, pest and disease management, irrigation, harvesting, yield, selling price and constraints were recorded using a structural questionnaire. The profitability analyses of potato and percentage share of each component were also calculated. Though the ground truth data on potato cultivation were not collected in 2003 the available data taken from farmers in the study area were used to compute the productivity in 2003.

3.2 Soil Properties

3.2.1 Sampling and soil preparation

Soil samples were taken randomly from 0 –20 cm depth from each experimental site covering hilly lands and lowlands (rich-lands) where potato has been grown. All samples were sieved through 2.0 mm sieve separately. Sub-samples of each soil were air dried and analysed for important properties.

3.2.2 Chemical properties of Soils

- Soil analysis for available nutrients

  Air dried soil samples (5.0 g) were weighed and mixed with 0.5N HCl and shaken overnight. Then the mixture was filtered into 100 ml volumetric flask and the resultant filtrate was used for analysis of N, P ,K, Ca and Mg. Total N was determined after adding 1:1 HCl to the filtrate and measuring photochemical absorbance using UV-Visible spectrophotometer. Total P was determined by colorimetric method with Ammonium Vandomolybdate. K, Ca and Mg were determined using an atomic absorption spectrophotometer.

- Organic carbon in Soils

  Modified Walkely and Black method was used to measure organic carbon of the soils.

- Soil pH was measured with a glass electrode using a soil to water ratio of 1 : 2.5.

3.2.3 Physical properties of soils

- Soil textural class

  Soil textural classes were determined using graphic guide for textural classification of less than 2.0 mm portion (Source: USDA – Soil Conservation Service).

- Soil colour

  Soil colour was described using the Munsel Colour Chart according to its Hue, Chroma and Value.

3.3 GIS Model

3.3.1 GIS Procedure

Available data

1 : 10,000 and 1:50,000 topographic maps published by Survey Department of Sri Lanka (1996)

Aerial photographs of 1: 20,000 (1999) from the Survey Department of Sri Lanka.

Geological map (1: 100,000), published by Geological Survey and Mines Bureau (1998).

Daily and monthly rainfall and temperature data from the Department of Meteorology, Sri Lanka.

Agroecological zones of Sri Lanka

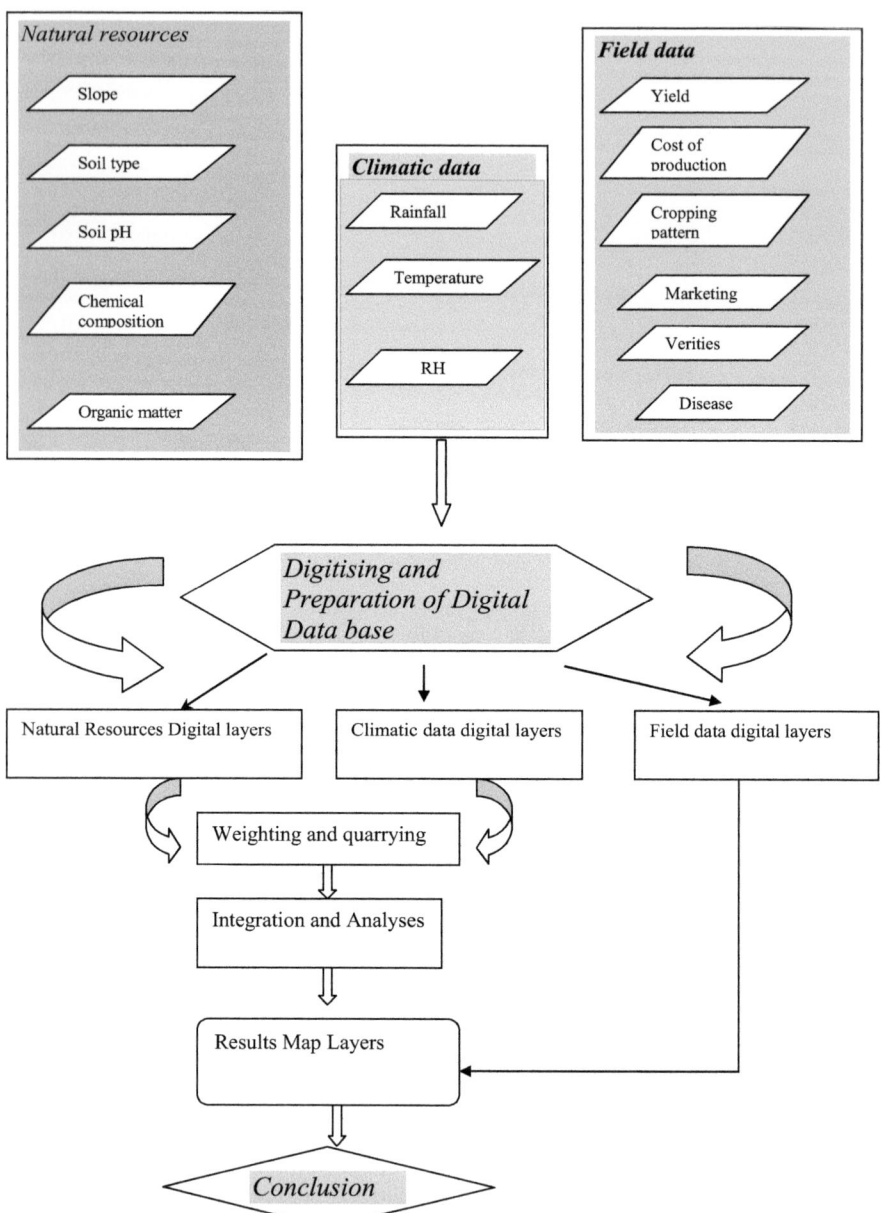

Figure 3.1. GIS Based Crop Production Model

Secondary Data.

    Soil Chemistry

    Soil pH

    RH

    Soil texture, types, moisture content

    Cultivation data (variety, cost, yield, disease, season)

Topographic maps of the study area were scanned and geo-referenced.

On screen digitising were under taken and following digital map layers were prepared

    Vector map of contour with attribute

    Vector map of Cluster boundary map (5 clusters)

    Roads map (vector)

    Drainage map (vector)

Aerial photographs were studied using standard Mirror stereoscope and demarcated potato growing areas.

Following map layers were prepared using processed secondary data

    Polygon map of rainfall

    Polygon map of benefit to cost ratio

    Polygon map of soil chemistry

    Raster map of slopes

    Polygon map of soil nutrients

Created suitable links between spatial and attribute data.

Carried out data analysis and overlay operations.

Performed weighting, rating and classifying for suitability analysis.

The four suitability aspects maps were combined into a composite suitability map by simple addition of the suitability weighting system. Weighting could be applied when not all aspects have an equal importance.

3.3.2 Weighting

pH

The optimum pH is about 5.5-6.0. Very acid soils make small potato tubers.

Score in map suit-score	Class	Class Code
< 5.0 & > 6.1	Unsuitable	1
5.0 –5.3	Moderately suitable	2
5.3 –5.5	Suitable	3
5.6 – 6.0	Highly suitable	4

Rainfall

Higher rainfall affects the tuber yield because it is favourable for occurrence and spread of late blight disease. Too much of water or too little will affect tuber number, size and quality.

Rainfall *"maha"* (mm)

Score in map suit-score	Class	Class Code
400 -500	Marginally suitable	2
500 -600	Highly Suitable	5
600 - 700	suitable	4
700-850	Moderately suitable	3
>850	Unsuitable	1

Rainfall "*yala*" (mm)

Score in map suit-score	Class	Class Code
< 120	Unsuitable	1
200 -300	Marginally suitable	2
300 -400	Suitable	3
>400	Highly suitable	4

Potassium (K) (ppm)

Potato is a heavy feeder in terms of potassium (Crop profile- Potato).

Score in map suit-score	Class	Class Code
<100	unsuitable	1
100-200	Marginally suitable	2
200-250	Moderately suitable	3
>250	Suitable	4

Phosphorus (P) (ppm)

Potato is a heavy feeder in terms of phosphorus (Crop profile – Potato) but phosphorus has an antagonistic effect on uptake of K by plants.

Score in map suit-score	Class	Class Code
1000-1450	Unsuitable	1
800-1000	Highly suitable	4
600- 800	suitable	3
500 - 600	Moderately suitable	2

Organic matter

Application of 10-12 t/ha of well- rotten cattle manure or compost in furrows before planting will improve growth and increase the yield.

Score in map suit-score %	Class	Class Code
< 1.0	Unsuitable	1
1.0 – 1.5	Moderately suitable	2
1.5– 3.0	Suitable	3
> 3.5	Highly suitable	4

Nitrogen (N)

Nitrogen must be carefully managed to provide adequate but not excess amounts. The usual ratio of carbon to nitrogen in the majority of cultivated soils is between 1:12. Narrow ratios are not common in most productive soils. Ratios above are common in nitrogen deficient soils (Allison, 1973).

C: N ratio

Score in map suit-score	Class	Class Code
<8	Marginally suitable	1
8-10	Moderately suitable	2
10-12	Highly suitable	4
>12	Suitable	3

Sol Textural Class

Potatoes grow best in deep to moderately deep, loose, well drained soils (William, 2001). Best soil is deep, well drained sandy or silty loam (Crop profile – Potato). Ill drained soils are generally unsuitable. Clay should be improved with organic matter.

Score in map suit-score	Class	Class Code
Loam	Suitable	3
Clay loam	Moderately suitable	2
Clay	Marginally suitable	1

4. RESULTS AND DISCUSSION

4.1 Location and Study Area

Nuwara Eliya and Badulla, major potato growing districts in Sri Lanka were selected for the study. Both districts located in Central hill country of Sri Lanka having general elevation over 1500 m. The area is characterised by having ridge and valley topography with steep slopes (Figures 4.1 and 4.2).

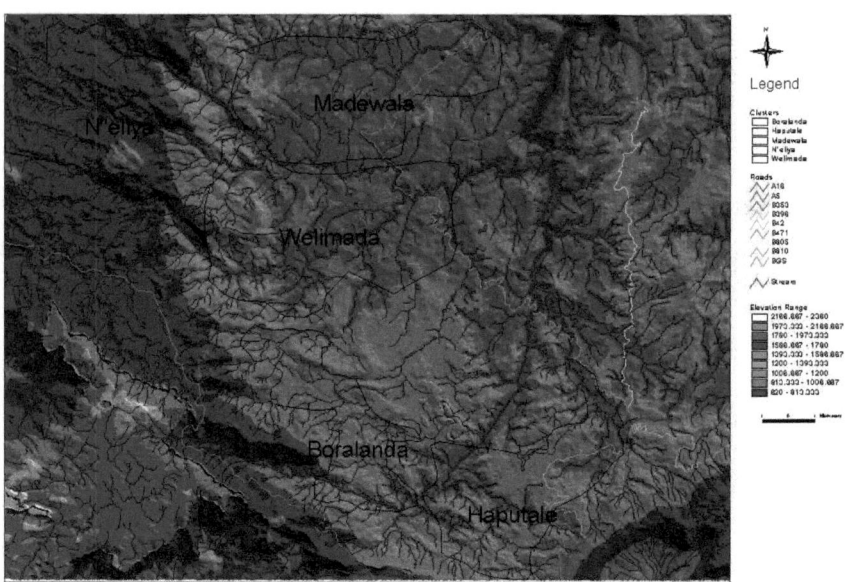

Figure 4.1. Digital elevation model of the study area

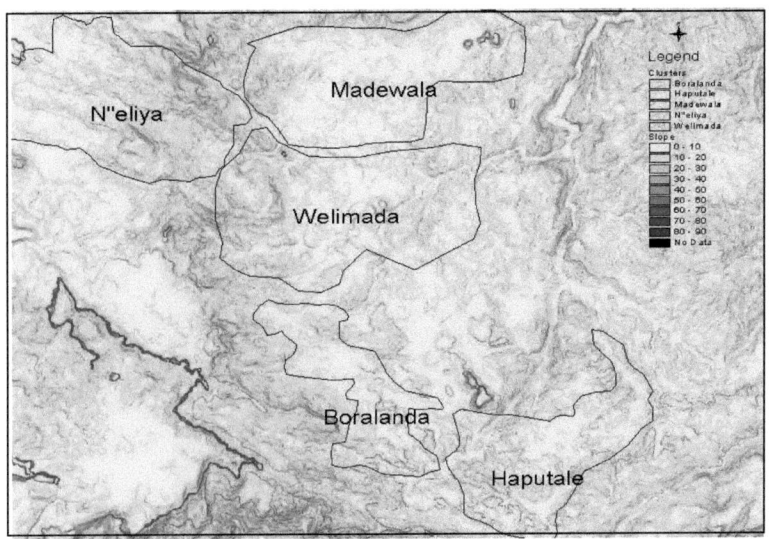

Figure 4.2. Slope map of the study area

4.2 Cropping System

In the surveys, it was observed that farmers in Nuwara Eliya grow potatoes in lowlands twice a year in both "*yala*" and "*maha*" seasons and in uplands only in "*maha*" season (Table 4.1). In Badulla district famers grow potatoes twice a year in uplands while one crop season in lowlands. In both districts, five potato grown clusters; Nuwara Eliya, Medawela, Welimada, Boralanda and Haputale were identified for the study (Figures 4.2).

Table 4.1. Cropping systems adopted by farmers.

District	Season	Land class	Cropping pattern	Irrigation
Nuwara Eliya	"*Yala*"	Lowland	Veg.-potato-veg-potato	Rain water
	"*Maha*"	Upland	Veg.-potato-veg.	Rain water

Badulla	"Yala"	Paddy fields	Paddy-potato-veg.	Irrigation
	"Maha"	Upland	veg.-potato-veg.	Rain water
	"Maha"	Upland	Fallow-potato	Rain water

*Veg.=Vegetables

Vegetables were grown between potato cultivations in Nuwara Eliya district. In Badulla district, farmers grew potatoes in lowlands once a year after harvesting rice and in uplands only in "*maha*" season. They grew vegetables in uplands before growing "*maha*" season potato crop. Major vegetable types grown in potato fields were beans, cabbage, carrots, leeks, beetroot, radish and salad leaves.

Better understanding of agronomic practices has been established as bean/cabbage - potato – bean/cabbage rotations in highland planting systems in Badulla district while rice – potato – rice rotation system was followed in lowlands. Hence, traditional potato farming system has been attempting to integrate potato and vegetable cultivation.

4.3 Planting and Harvesting

Usually, potato is propagated vegetatively. The shoots develop from the "eyes" of a "seed" tuber. The new tuber is formed at the tips of underground stolons which are adventitious shoots formed at the base of the stem. Number of shoots emerges from the seed piece and they grow together forming multi-shooted plants (Plates 4.1 and 4.2). The vegetative propagation makes potato cultivation particularly vulnerable to diseases because in contrast to propagation by true seeds, infected plants will transmit the disease to the next tuber generation.

Time taken for maturity of potato tubers vary and depends on the variety being grown. Usually, dying of shoots indicates the maturity of tubers. Tubers were harvested by gently digging them up with an agricultural tool and gently uprooting the plant (Plate 4.3).

4.4 Varieties Grown in Nuwara Eliya and Badulla Districts

Though the Department of Agriculture has recommended number of potato varieties for cultivation in Sri Lanka, the main commercial varieties being Granola, Desiree, Lyra, Raja, Binella and Arnova. Granola was the popular one among farmers because its potential for high yielding and relatively short time for tuber maturity despite relatively low resistance to late blight disease.

Plates 4.1. One month old potato plantation in Nuwra Eliya

Plate 4.2. Eight week old potato plantation

Plate 4.3. Harvesting upland grown potatoes

The average yield performance of the potato varieties grown in *"yala"* and *"maha"* seasons over three years (2003 – 2005) at five clusters were tabulated separately (Tables 4.2 a and 4.2b).

Table 4.2a. Yield performance of potato varieties during "*yala*" season
    in 5 clusters (t/ha).

	Yala				
Variety	N'Eliya	Welimada	Boralanda	Haputhle	Medawela
Granola	24.0	23.4	28.1	20.0	21.0
Lyra	20.0	16.3	-	-	--
Raja	-	-	-	-	-
Desire	-	15.0	-		-
Arnova	-	21.3	-	-	-
Binella	-	-	27.5	-	-
Ricolta	-	-	-	18.8	-

Table 4.2b. Yield performance of potato varieties during "*maha*" season
    in 5 clusters (t/ha).

	"Maha"				
Variety	N'Eliya	Welimada	Boralanda	Haputale	Medawela
Granola	20.4	21.4	21.5	20.6	13.3
Lyra	-	20.3	-	-	-
Raja	15.0	-		-	-
Desire	-	21.7	-	20.0	12.5
Arnova	-	-	-	22.5	-
Binella	-	-	17.5	7.5	-

The highest yield was recorded by Granola in all clusters in "*yala*" season while
Binella showed the second best performance. In "*maha*" season Granola recorded
the high yield in four clusters except Medawela while Desiree and Arnova recorded

43

the high yield in Welimada and Haputale. It was also observed that there was no remarkable yield difference among the high yielding three varieties. Granola, Desiree and Arnova performed well under rain-fed conditions. The variety showing the highest average tuber yield was selected as the most suitable variety for the location and the season.

4.5 Soil Chemical Properties

It is evident from the results that there is no significant difference between lowland and upland soils in terms of chemical properties in different clusters (Figure 4.3).

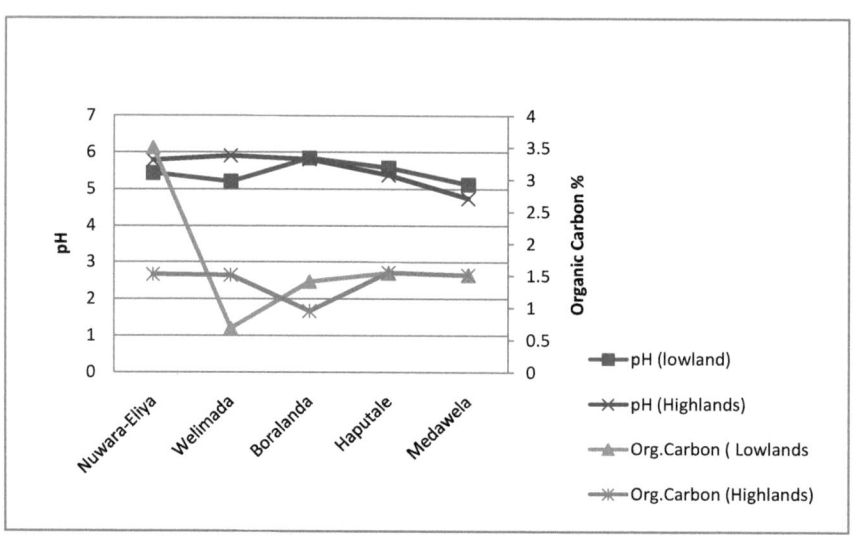

Figure 4.3. Mean soil pH and organic carbon in different clusters.

All soils were in acidic reaction and low in organic matter except Nuwara Eliya lowlands. Results also showed that all soils were in suitable pH range for potato cultivation (5 –6) except Medawela uplands which had soil pH less than 5.

44

4.6 Soil Physical Properties

Most of upland soils were loam except a few sites whereas majority of lowland soils was clay. It may be because almost all potato grown lowlands in Badulla district were paddy fields (Table 4.3).

4.7 Rainfall Pattern

Rainfall pattern of the years showed that heavy rainfall experienced during *"maha"* season and rainfall intensity was remarkably higher than that of *"yala"* season in the same year (Figures 4.4 and 4.5). *"Maha"* season is a wet season with a weekly average of 4 –5 rainy days (Babu, 2005).

Table 4.3 Soil textural classes and soil colours

Highland			
Cluster	Site	Textural class	Colour
N'Eliya	1	Loam	Light brown (7.5YR 6/3) dry
	2	Clay loam	Reddish brown (5YR 6/3) dry
	3	Loam	Reddish brown (5YR 3/2) dry
Welimada	1	Loam	Pale brown (10YR 6/3) dry
	2	Loam	Ash brown (2.5YR 4/2) dry
	3	Loam	Pinkish brown (10YR 6/3) dry
Boralanda	1	Loam	Pinkish gray (7.5YR 6/2) dry
	2	Loam	Yellowish brown (10YR 4/4) wet
	3	Loam	Pale brown (10YR 6/2) dry
Haputale	1	Clay loam	Yellowish brown (10YR 3/4) wet
	2	Silty clay	Brown (2.5YR 4/2) dry
	3	Silty clay	Yellowish brown (10YR 3/3) dry
Medawela	1	Loam	Reddish brown (5YR 4/4) dry
	2	Loam	Pinkish brown (5YR 3/4) dry
	3	Clay loam	Reddish brown (2.5YR 43/4) dry

Lowland		
Site	Textural class	Colour
1	Clay loam	Reddish black (10YR 2.5/1) dry
2	Clay	Reddish brown (5YR 3/2) dry
3	clay	Reddish brown (5YR 3/2) dry
1	Clay	Reddish brown (5YR 4/4) dry
2	Clay loam	Pinkish brown (2.5YR 5/4) wet
3	Clay	Ash (2.5YR 4/2) wet
1	Clay	Reddish brown (5YR 4/4) dry
2	Clay	Blackish brown (2.5YR 3/1) dry
3	Clay loam	Reddish brown (5YR 3/4) wet
1	Clay	Brown (10YR 3/4) wet
2	Clay	Blackish brown (2.5YR 3/1) dry
3	Clay	Brown (10YR 3/4) wet
1	Clay	Blacish brown (2.5YR 3/1) dry
2	Clay	Ash brown (7.5YR 4/2) dry
3	Clay	Reddish brown (5YR 4/4) wet

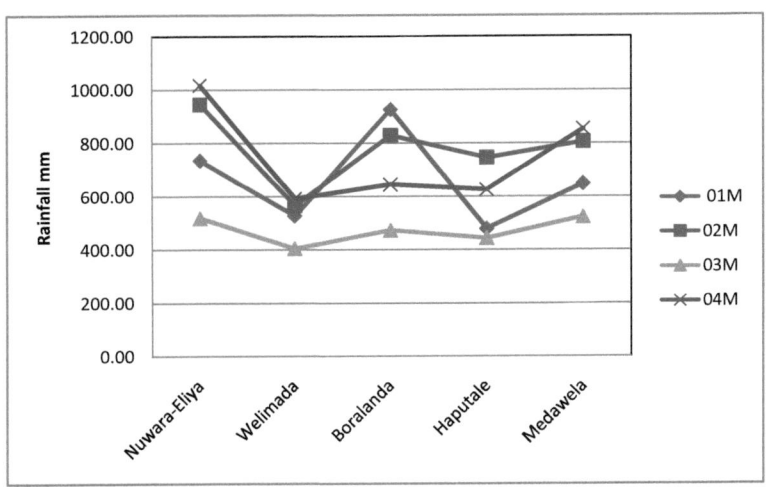

Figure 4.4. Rainfall over "*maha*" seasons in different clusters (2001-2004)

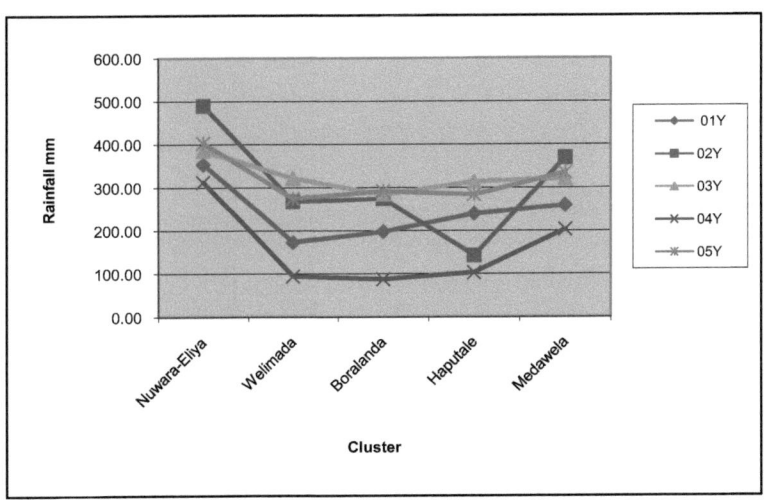

Figure 4.5 Rainfall over "*yala*" seasons in different clusters (2001-2005)

4.8 Irrigation and Fertilization

Water supply is another critical factor that limits the extent of potato cultivation to the up-country. In hills where potatoes are planted in *"maha"* season it is difficult to regulate moisture content where potatoes have to be grown as a rain-fed crop. Research and agronomic practices have revealed that irrigation at the time of tuber initiation leads to an appreciable improvement in the potato yield.

Fertilizer is another expensive input, second only to seed cost in potato cultivation. Optimum fertilizer application for maximum yields under different agroecological conditions will help farmers tremendously to reduce cost of production, which is the key factor for net income.

Management of water and nitrogen is particularly challenging in potatoes. Under dry conditions plants become yellow and grow slowly. Although these plants may appear N-deficient, adding nitrogen to drought-stressed potatoes worsens the problem by causing salt damage and allowing soil- borne pathogens such as *Fusarium* spp. to enter damaged seed pieces. Even where salts are not a problem, high levels of nitrogen and water can lead to hollow heart in large tubers because of rapid tuber expansion. Small, but adequate nitrogen applications, spread out over long period will reduce the potential for hollow heart and brown centre, a related physiological disorder. Excess nitrogen especially, late in the growing season decreases dry matter content resulting in poor storage quality and poor texture when cooked.

4.9 Productivity of Potato

Fresh tuber yield performance of the potato varied with seasons (Figure 4.6). The *"yala"* season being relatively drier than the *"maha"* season produced a remarkable higher potato tuber yield than that of wet *"maha"* season at every cluster irrespective of soil nutrient levels. Furthermore, the yield performance of *"yala'* season is always greater than that of *"maha"* season in the same year during the study period. It was also observed that all commercial varieties were susceptible to late blight and wilt diseases during '*maha*" season, which could be attributed to decrease in productivity (Figures 4.7).

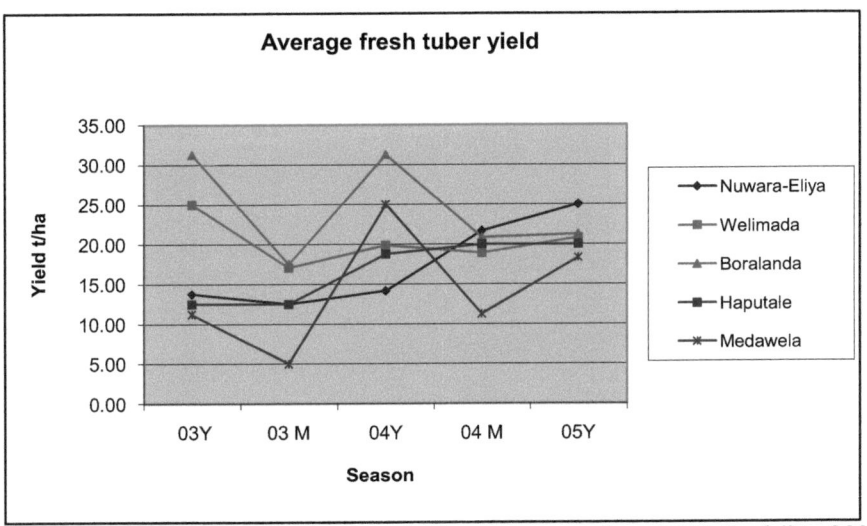

Figure 4.6. Average fresh tuber yield over seasons from 2003 *"yala"* to 2005 *"yala"* season.

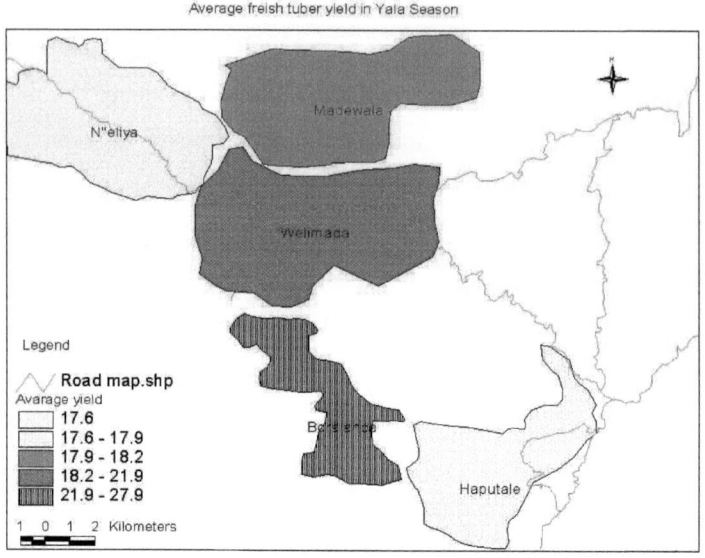

Figure 4.7. Average fresh tuber yield in *"yala"* season (2003-2005)

4.10 Occurrence of Diseases

Following diseases were observed in both districts during the study period.

Bacterial wilt of potato caused by *Ralstonia solanacearum*

Early blight of potato caused by *Alternaria solani*

Late blight of potato caused by *Phytophthora infestans* (Plate 4.4)

Plate 4.4. Potato leaf showing late blight symptoms

Most farmers timely applied recommended fungicides and protectants and managed to minimise disease level to keep the loss below an economic injury level. Meanwhile some farmers were unable to protect their crop from late blight spread due to heavy rain, which was beyond farmers' control. Bacterial wilt was reported in upland grown crops in both districts at different damage levels. Some potato fields were badly affected by wilt and caused remarkable yield loss. However, it was also observed that lowland (paddy field) grown potatoes were free from bacterial wilt, and late blight was the main destructive disease in lowland (paddy field) grown potatoes (Table 4.4).

Further, it was observed that all commercial varieties were susceptible to late blight and wilt during "*maha*" season, which could be attributed to decease in productivity.

On an average Granola fresh tuber yield remained 20 –24 t/ha according to the recommendation given for Sri Lanka, *i.e.* 8-12 kg of marketable tubers from each kg of seed planted. The diseases have severely affected tuber yield, which in turn reduced the net profit to the farmer.

Table 4.4. Effect of diseases on fresh tuber yield (t/ha)

Village	Variety	Diseases	Seed: Harvest	Land use
Katumana	Granola	Late blight	01:05	Carrot-potato
Suwadelpola	Raja	Wilt and seed rot	01:02	Bean - potato
Dikkapitiya	Granola	Tuber rot	01:05	Fallow - potato
Mirahawaththa	Granola	Late blight, wilt	01:05	Cabbage potato
Kalubululanda	Binella	Late blight	01:06	Radish-potato
Haputale	Desiree	Late blight .wilt	01:03	Bean-potato
Uwaparanagama	Granola	Late blight, wilt	01:04.5	Fallow- potato
Uwaparanagama	Desiree	Late blight	01:04	Paddy-potato
Uwaparanagama	Granola	Late blight, seed rot	01:02	Paddy-potato

In Badulla, "*yala*" season potatoes are planted in flood irrigated paddy fields in which most of the wilt causing bacteria in the soils are killed due to anaerobic condition and most potato crops are free from wilt (*R. solanacearum*) disease. Babu (1999) reported that 21.8% of the total potato crop is lost by potato diseases. Raja, Escot, Krushi, Sita and Maranka posses high level of late blight tolerance while Hillstar possesses high level of late blight resistance. Since farmers used to purchase seed potatoes whatever available in the market during planting time they do not have option to get suitable disease resistant varieties at planting time.

4.11 Marketing

Majority of farmers sell their produce to Pettah, Haputale and Keppetipola markets and a few sell to private dealers at farm gate. The selling prices were recorded separately for 5 different clusters. The selling prices fluctuated widely throughout the year.

4.12 Cost of Cultivation

The cost of cultivation was calculated in each season in each cluster using expenditure over potato growing period excluding labour inputs. Since labour inputs were shared among family members of farmers at different levels, labour cost cannot be estimated.

Cost of cultivation of potato in main growing areas in year 2004 *"yala"* and *"maha"* is shown in Table 4.5. When considering the different cost components, seed potato is still accounting for more than 55%. The cost of producing 1.0 kg of potato during *"yala"* in Nuwara Eliya and Badulla districts was Rs. 25.13 and Rs. 22.60 respectively whereas in *"maha"* Nuwara Eliya and Badulla districts account for Rs. 23.09 and 23.75 respectively.

Table 4.5. Cost of production of potato in Nuwar Eliya and Haputale (Badulla) clusters in 2004.

Cost component (for 50 kg of seed)	2004 "*Yala*"				2004 "*Maha*"			
	Nuwara Eliya		Badulla		Nuwara Eliya		Badulla	
	Rs.	%	Rs.	%	Rs.	%	Rs.	%
Seed	6200	61 69	5500	64.89	5500	55.0	6000	63.15
Fertilizer	1500	14.92	1300	15.33	1300	17.0	1700	17.89
Pest & disease control	1500	14.92	1000	11.79	1500	15.0	1000	10.52
Miscellaneous	800	7.96	675	7.96	1300	13.0	800	8.42
Total cost	10050		8475		10000		9500	
Average yield (kg)	400		375		433		400	
Unit cost/kg	25.13		22.6		23.09		23.75	

The income from potato cultivation was calculated multiplying average price and average production. The income and cost data for seasons from 2003 to 2005 were used to calculate benefit to cost ratio of each cluster (Figure 4.8). These data were used to study seasonal price fluctuations and yearly fluctuations. Results reveal that the average productivity of potato in Badulla and Nuwara Eliya districts was 22 t/ha and it is relatively low compared to other potato growing countries in the world and the potential yield of the varieties. The cost of production is very high (nearly Rs 22-25/kg) since seed potatoes are very expensive (Rs.110-124/kg) and accounts for more than 55% of the total production cost (Table 4.5). However, seed potato price vary with ware-potato prices. When ware-potato prices are high seed potato prices are also high (Figures 4.8, 4.9 and 4.10).

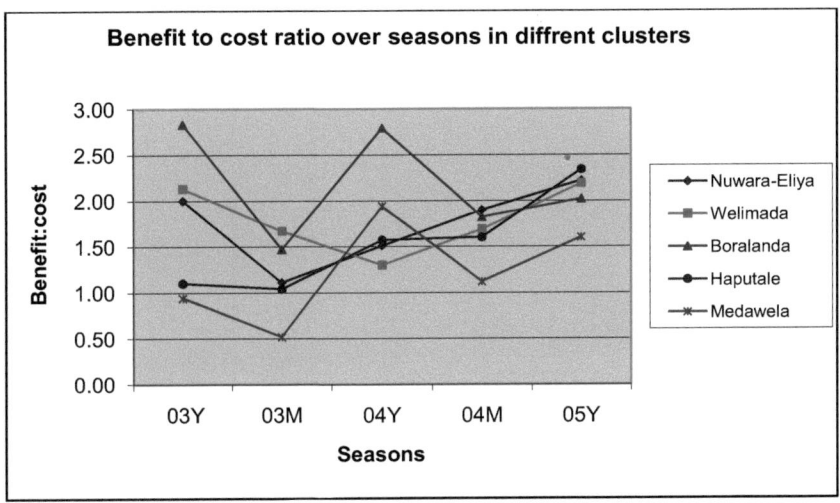

Figure 4.8. Benefit to cost ratio over seasons in different clusters.

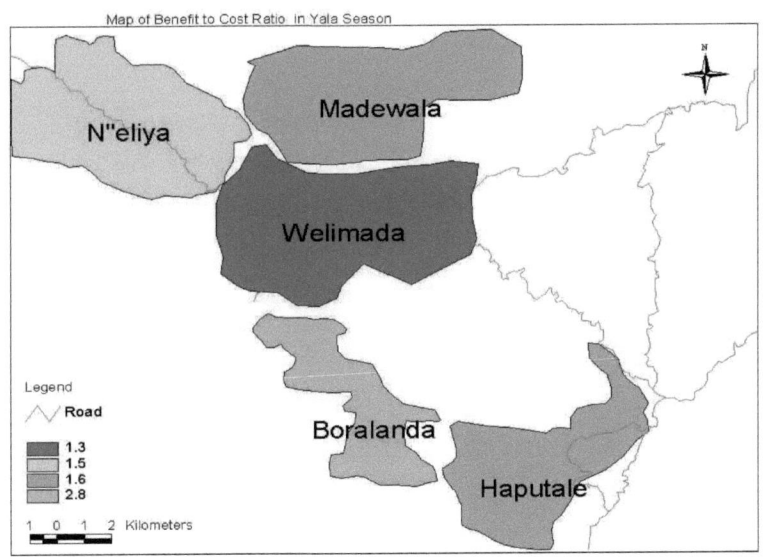

Figure 4.9. Map of benefit to cost ratio in "*yala*" season

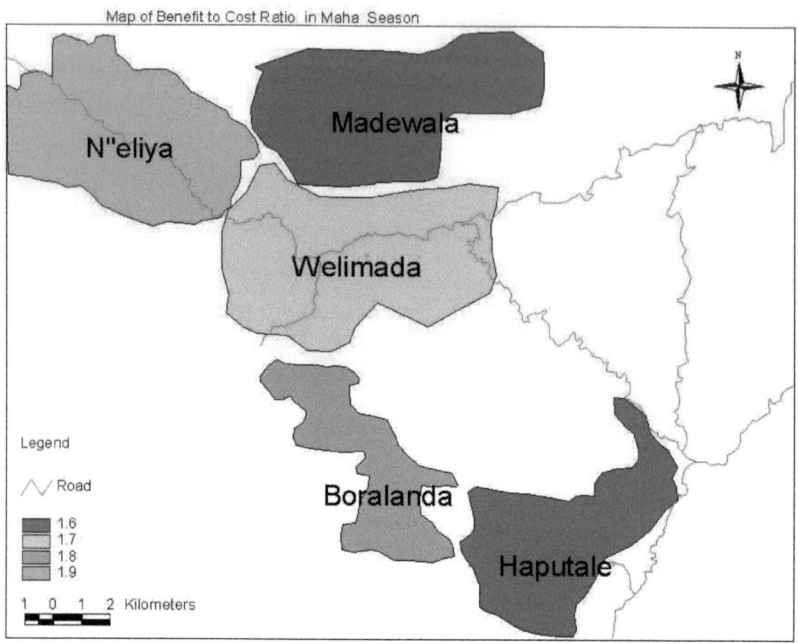

Figure 4.10. Map of benefit to cost ratio in "*maha*" season

To compare the average tuber yield and composite weights, average yield of three years (2003-2005) in different clusters were computed (Figure 4.11). Results reveal that in "*yala*" season of 2003 and 2004, Boralanda recorded the highest benefit to cost ratio due to the promising yield received in that season (Figures 4.6 and 4.7).

56

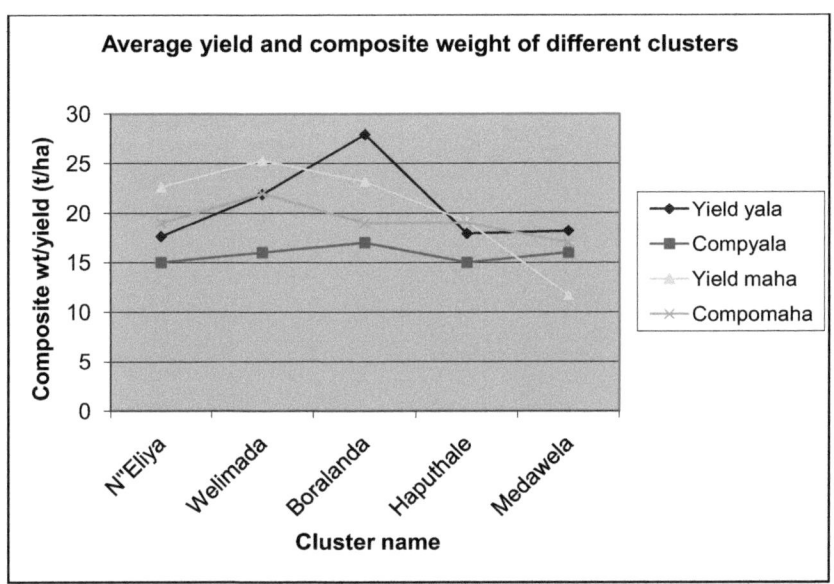

Figure 4.11. Composite weights and average tuber yield in different clusters

Use of GIS for modelling of potato cultivation in Sri Lanka is a fairly complicated exercise, because the parameters, which affect the yield very often, do not show linear relations. Available literature and our field information were used to classify and give weights for the parameters. Because of heterogeneous nature of soil, climatic conditions and all other physical parameters, modelling of the productivity and land suitability is not an easy task. However, with the help of the statistical analysis the data were generalized. Therefore, some deviations from the ground truth data can be observed.

Currently, a wide range of new varieties is being assessed for suitability for cultivation in up-country (Babu *et al.*, 2005). According to the survey, the main variety cultivated during the study period was Granola. It is extensively cultivated because it is an early maturing (60-75 days), high yielding variety and produces a high proportion of large-sized tubers. However, the main reason for its popularity is due to slow rate of seed degeneration and, due in part to its resistance to potato cyst

57

nematode (Department of Agriculture, Sri Lanka, 2006). Though Granola was popular among farmers it is susceptible to late blight and wilt diseases. Further, Granola and Binella in Boralanda cluster were comparable in terms of yield, showing their suitability to up-country intermediate zone for "*yala*" season in lowland cultivation whereas Granola, Desiree and Arnova are suitable for "*maha*" season in upland cultivation.

However, the productivity is also constrained by inappropriate varieties, especially for susceptibility to diseases, the high cost of planting material, fertilizer and pesticides, inefficient farming practices and poor infrastructure especially transport which increase production and marketing costs.

Further, the results of the present study show that majority of farmers have small land holdings. On an average the orchard area of a potato was found less than 0.2 ha. More than 80% farmers have decreased the area under potato crop during last few years. More than 70% farmers were not satisfied with the quality of the seed they received from the market. The main source of irrigation is canals and rain water. Most farmers in the study area applied chemical fertilizer (NPK) only while some farmers applied poultry manure along with chemical fertilizer. Scarcity of irrigation water is the biggest constrain in upland cultivation followed by non-availability of inputs especially certified good quality seed.

Constrains

i) High seed cost; seed cost itself contributes to $> 55\%$ of total cost of production and small scale farmers are not in a position to afford.

ii) Inadequate availability of good quality seed and lack of regular supplying of high quality seed at correct time.

iii) Farmers experienced that some imported seed potatoes were contaminated with pest and diseases, and poor quality too.

iv) Use of imbalanced fertilizer.

v) High incidence of pests and diseases and improper pesticide application.

vi) High degree of dependence on the rain-fed system.

vii) Inefficient resource management practices adopted by farmers leading to low productivity, resulting in high cost of production.

viii) Seasonality of production and drastic price fluctuation.

ix) Inadequate marketing facilities and poorly operated distribution network.

x) Unavailability of support services for small scale farmers and lack of farmer awareness programmes.

xi) There is a shortage of high yielding pest and disease resistant varities.

xii) Improper crop management strategies leading to high cost of production and low profit margin.

5. CONCLUSIONS AND RECOMMENDATIONS

Results of the present study revealed that Granola produced high yields constantly for the six seasons throughout the study period under climatic conditions of central hills of Sri Lanka,

The cost of potato production has increased tremendously during the past few years and consequently potato production has become uneconomical for small scale farmers, especially in Badulla district. Therefore, more lands have moved away from potato to vegetable crops. The attractive price of vegetable crops, which expanded the margin of profit, encourages the farmers to grow vegetable crops in potato lands. Presently, part of the extent has been diversified into vegetables. The cost of production has increased with the reduction in fertilizer subsidy and the other direct and indirect subsidies.

According to the composite scores in GIS model the highest score was recorded in Welmada in "*maha*" season. This is because Welimada area receives favouble rainfall, possesses suitable soil pH, maintains adequate amount of organic matter and nitrogen during the growing seasons. However, Medawela cluster recorded the lowest value due to deviation from optimum soil pH and lack of available P and K to the plant.

In "*yala*" season the highest composite score was recorded in Boralanda which is compatible with benefit to cost ratio of Borlanda. Haputale and Nuwara Eliya received lower scores due to excess nitrogen and calcium, which affect the tuber yield.

In both "*yala*" and "*maha*" seasons, composite scores and average fresh tuber yield were positively correlated. It is evident from the results the average highest yield during three year period (2003-2005) was recorded in Boralanda and Welmada clusters in "*yala*" and "*maha*" seasons respectively.

The crop yield resulting from luxuriant growth will depend on irrigation and fertilizer used in the cropping system, and these in turn will vary from field to field and farmer to farmer. Further, the crop climate may vary accordingly among the fields. The above possibilities imply that crop yield vary from field to field within the same cluster. Therefore, a generalised productivity model for the entire locality may not give benefit to all categories of farmers in the potato growing areas.

Since most farmers in Nuwara Eliya and Badulla districts are small scale farmers their performance data are mostly heterogeneous. Further, the field data vary from season to season in different clusters. It would be more accurate if number of sites and seasons increased for computation.

Parameters used in GIS model and scoring system are mostly based on our research findings and available literature, but there are many more factors to be considered such as prominent climatic factors and soil properties.

Benefit to cost ratio can be increased by reducing cost of production providing incentives as well as establishing proper marketing system.

The yield and net return to the grower can be increased by providing disease resistant good quality seed on time.

Finally, no doubt, potato is a profitable enterprise and can compete with other crops of the season in terms of the profit.

ACKNOWLEDGEMENTS

We wish to thank University of Sri Jayewardenepura for providing financial assistance (Grant No. ASP/06/RE/2003/04) to carry out this research project. We would like to express our appreciation to Mr. W.M.K.S Weerasekera, Technical Assistant for helping to collect field data. We also appreciate help given by Mrs.Shamali Siriwardena, Chief Chemist, Geological Survey and Mines Bureau for assisting in chemical analysis of soil data and Miss. Wathsala for assisting to prepare maps.

REFERENCES

Abeyaratna, K.W.S.D. (1985). Potato production in Sri Lanka. International Potato course. IAC, Wageningen.

Angammana,V.S.(1983). Report on potato production in Sri Lanka. International Potato Course IAC, Wageningen.

Annon (1993). Assessment: Potatoes in Sri Lanka.

Allison, F.E.(1973). *Soil organic matter and its role in crop production*. Elseveir Scientific Publishing Company, New York.

Babu, A.G.C. (1999). Important diseases of potato and their management. Potato Report. Department of Agriculture.

Kularatne, R.J.K.N. (2003). Potatoes and Seed Potatoes Production in Sri Lanka. Potato Report. Department of Agriculture.

Mahakumbura, L.M.M.(1980). Report on potato production in Sri Lanka. International Potato Course IAC, Wageningen.

Mazeen, A.C.M., Abeykoon, S.A.M.R., Nugaliyadde, M.M., Marambe, B., Kularatne, R.J.K.N. and Amarasena,B.G. (2002). Effect of age of mother plants on growth, yield and senescence of rooted stem cuttings of basic seed potato. *Annals of the Sri Lanka Department of Agriculture*,4,161-168.

Nugaliyadde, M.M (2003). Strengthening of informal seed potato production systems in Sri Lanka. Briefing Paper, Law & Society Trust, Sri Lanka.

Panabokke, C.R. (1996). Soils and Agro-ecological environments in Sri Lanka. Natural Resource Service No.2. Natural resource and Science authority of Sri Lanka.47/5, Maitland Place, Colombo 07, Sri Lanka.

Patummintra, S. and Kesawapitak, P. (2002). The monthly production potential in the eastern province of Thailand by using the rubber production model and geoinfomatics. WCSS, Thailand.

Sathiamoothy, K., Prange,R., Mapplebeck,L. and Haliburton,T. (1985). Potato production in Sri Lanka. *American Potato Journal*, **62**(10): 555-564.

Crop Profile – Potato
http:/www.cals.ncsu.edu/sustainable/peet/profiles/pppotato.html

Department of Agriculture, Sri Lanka –www.agridept.gov.lk/ndex.php/en/crop

State-wide IPM program, Agriculture and Natural Resources, University of California (2003). http://www.ipm.ucdavis.edu/PGM/selectnewpest.potatoes.html

William, J.L. (2001). Agricultural http:/agalternatives.aers.psu.edu.

yes

I want morebooks!

Buy your books fast and straightforward online - at one of the world's fastest growing online book stores! Environmentally sound due to Print-on-Demand technologies.

Buy your books online at
www.get-morebooks.com

Kaufen Sie Ihre Bücher schnell und unkompliziert online – auf einer der am schnellsten wachsenden Buchhandelsplattformen weltweit!
Dank Print-On-Demand umwelt- und ressourcenschonend produziert.

Bücher schneller online kaufen
www.morebooks.de

OmniScriptum Marketing DEU GmbH
Heinrich-Böcking-Str. 6-8
D - 66121 Saarbrücken
Telefax: +49 681 93 81 567-9

info@omniscriptum.com
www.omniscriptum.com

MIX
Papier aus verantwortungsvollen Quellen
Paper from responsible sources
FSC® C105338

Printed by Books on Demand GmbH, Norderstedt / Germany